# THE WILLOUGHBY GARDNER LIBRARY

*A collection of early printed books on natural history*

A selection of books from the collection

# THE WILLOUGHBY GARDNER LIBRARY

*A collection of early printed books
on natural history*

by
John R. Kenyon

Amgueddfa Genedlaethol Cymru
National Museum of Wales
Cardiff 1982

Cover: The heron from Belon's *L'histoire de la nature des oyseaux*, Paris, 1555

# FOREWORD

A library is an integral part of a museum in that it provides an essential
service for the curatorial staff in their research work, in their curatorial
duties and in their educational activities. Yet it is probably true that
few people outside the museum service are aware of the existence of
such libraries let alone the extensive nature of their holdings and the
important collections that they contain.

The Library of the National Museum of Wales is no exception.
Many of its items, especially in the field of natural history, have been
donated. One of the most notable of such gifts is the Willoughby
Gardner Collection which has recently been catalogued by John
Kenyon, the Assistant Librarian. Among Gardner's varied interests was
the development of natural history as a science and this led him to
collect several of the seminal printed works of the sixteenth,
seventeenth and eighteenth centuries.

The Gardner library was acquired by the Museum in 1953 and was
housed for several years in the Department of Zoology before being
transferred to the main Library. A selection from the bequest was
chosen for a special exhibition in the Main Building of the Museum in
1981 to mark the fifth centenary of the printing of the earliest book in
the collection, and later at Oriel Eryri, Llanberis, the Museum's newest
outstation, in Summer 1982.

It is fitting that this catalogue of Gardner's library initiates a series
of catalogues of the Library's holdings, and particularly as it is being
published during the 75th Anniversary of the granting of the Royal
Charter of Incorporation to the Museum.

It is intended that further important material in the Library will
receive similar treatment, the priority (as far as a published catalogue is
concerned) lying with the very important and large library on Mollusca
which belonged to J.R. le B. Tomlin. Many of these items were given to
the Museum by Tomlin in his lifetime as a result of his close association
with the institution, with the remainder coming after his death in
1954.

DOUGLAS A. BASSETT
Director
July 1982

# Introduction

The volumes which are the subject of this catalogue were bequeathed to the National Museum of Wales by Willoughby Gardner DSc., FLS, FRES, FSA (1860 – 1953), and came to the Museum in 1953. The majority of the library consists of early printed books on the natural sciences, ranging in date from the fifteenth century to the eighteenth century, and includes a selection of modern texts and commentaries. The library is kept as an entity in the Department of Zoology as most of the books relate to this field, but other works include herbals by Brunfels, Dodoens, Fuchs and Gerard as well as Gesner's work on fossils and there are plans to move the collection to the main library.

Gardner lived for much of his life in Deganwy, North Wales, and was a man who had varied interests, with publications on archaeology, entomology and numismatics. His interest in entomology, particularly bees, wasps, and butterflies, is evident from the books in his collection, including works by Aldrovandi, Merian, Sepp and Harris as well as several early treatises on bees and bee-keeping; often he obtained two or three copies of the same book. Shortly before his death he donated to the museum his collection of Aculeate Hymenoptera of North Wales containing 2,289 specimens.

Unfortunately, the library does not contain all the early books once owned by Gardner. At some date he presented to the Royal Entomological Society a copy of Pena and L'Obel's *Nova stirpium adversaria* (1576) that was once owned by Thomas Moffet, and which the Society has since sold.

A full obituary of Gardner was published in *The North Western Naturalist*, N.S.2 (1954 – 5), 652 – 55.

The catalogue entries have been kept concise in order that the catalogue could be produced as soon as possible, replacing the simple typed list which was inadequate both for security and for bibliographic purposes. A precise collation of one or two of the volumes must await a comparison with copies in other institutions, for example Gesner's *De omni rerum fossilium*, and I would be grateful if any errors or matters arising from the catalogue could be brought to my attention.

J. R. KENYON
Assistant Librarian
1982

**Aelianus,** Claudius [fl. 3rd century]

Historia per petrum gyllium latini facti . . . libri XVI. De ui & natura animalium.

Lugduni: Apud Seb. Gryphium, 1533, [26], 598, [2] pp., 22 cm. (4to)

**Aelianus,** Claudius

De historia animalium libri XVII.

Lugduni: Apud Guliel. Rouillium, 1565, [16], 668, [37] pp., 17cm. (8vo)

**Aelianus,** Claudius

De animalium natura libri XVII.

Genevae: Apud Philippum Albertum, 1616, [8], 1018, [94] pp., 11 cm. (8vo)

**Aelianus,** Claudius

De natura animalium libri XVII. Cum animadversionibus Conradi Gesneri, et Danielis Wilhelmi Trilleri: curante Abrahamo Gronovio.

Basileae: Apud Joh. Ludov. Brandmullerum, 1750, xii, lxviii, 1128, [90] pp., 21 cm. (4to)

**Albertus Magnus** [c. 1200 – 80]

De animalibus libri XXVI, [ed. by] Hermann Stadler.

Münster: Aschendorffsche, 1916 – 20, 2 v., 24 cm.

**Albin,** Eleazar [fl. 1713 – 59]

A natural history of English insects.

London: Printed for the author, 1720, [112] pp., single sided, 3 page handwritten index, 100 pls., 30 cm. (4to)

**Albin,** Eleazar

A natural history of English insects.

[50 out of the 100 plates which were published as a volume in 1720; the handwritten descriptions of the plates are taken from the published volume]

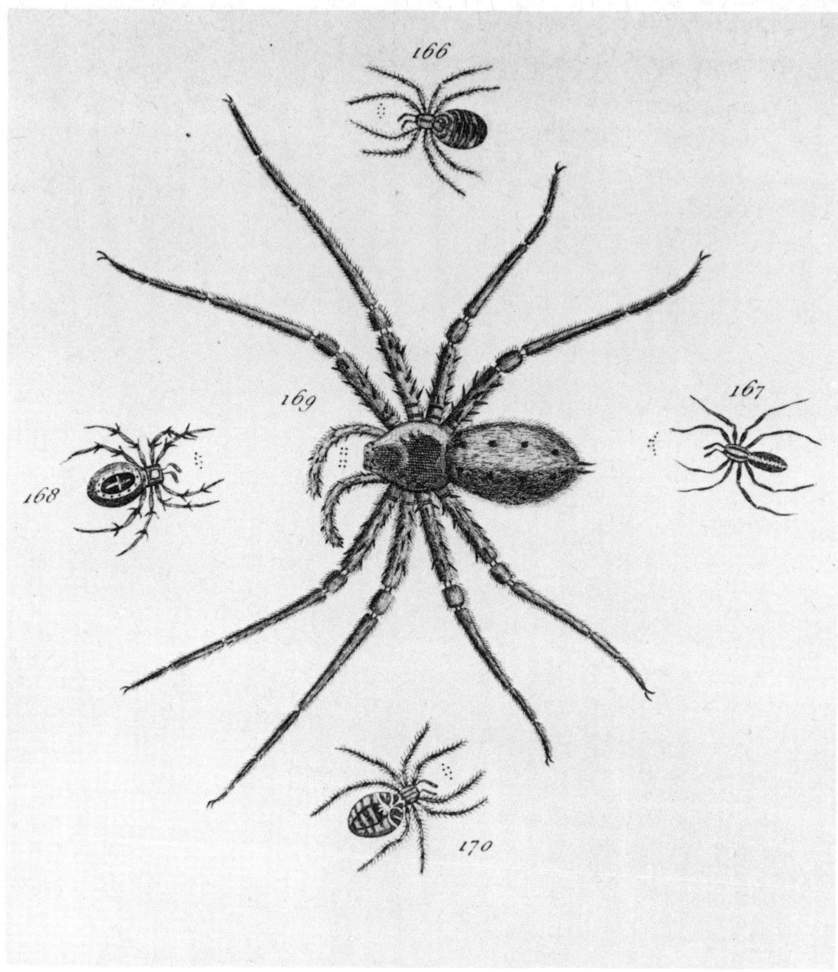

A plate from Albin's *A natural history of spiders,* London, 1736

## **Albin,** Eleazar

A natural history of English insects. To which are added, large notes, and many curious observations by W. Derham.

London: Printed by William and John Innys, 1724, [4], 28, [112] pp., single sided, 100 pls. (col.), 30 cm. (4to)


## **Albin,** Eleazar

A natural history of spiders and other curious insects.

London: Printed by John Tilly, 1736, [10], 76 pp., front., 53 pls., 29 cm. (4to)

**Aldrovandi,** Ulisse [1522 – 1605]

Ornithologiae hoc est de avibus historiae libri XII.

Bononiae: Apud Io. Baptistam Bellagambam, 1599, 1603 (vols. 1 & 3); Apud Nicolaum Tebaldinum, 1634, (vol. 2), 3 v., illus., 35 cm. (fol.)


**Aldrovandi,** Ulisse

De animalibus insectis libri septem, cum singulorum iconibus ad vivum expressis.

Bonon: apud Ioan. Bapt. Bellagambam, 1602, [10], 767, [41] pp., front., illus., 37 cm. (fol.)


**Aldrovandi,** Ulisse

De animalibus insectis libri septem, cum singulorum iconibus ad vivum expressis.

Bonon: Apud Clementem Ferronium, 1638, [10], 767, [43] pp., illus., 35, 36, 38 cm. (fol.)

[Colophon date 1644. End papers of one copy consist of an English tract and broadside dated 1641]


**Aldrovandi,** Ulisse

Memorie della vita di Ulisse Aldrovandi, medico e filosofo Bolognese.

Bologna: Lelio dalla Volpe, 1774, vi, 264 pp., front., 21 cm. (8vo)


**Alemann,** Conrad von, *of Magdeburg*

Das Buch der Natur, [ed. by] Hugo Schulz.

Greifswald: Abel, 1897, x, 445 pp., 1 pl. [inserted], 25 cm.


**Aristotle,** [384 – 322 B.C.]

De natura animalium.

[sl.: s.n.], n.d., [12], 328 leaves, 17 cm. (8vo)


**Aristotle**

De historia animalium lib. IX. De partibus animalium, et earum causis libri IIII. De generatione animalium libri V. Theodoro Gaza Thessalonicesi interprete.

Venetiis: apud Hieronymum Scotum, 1545, [8], 321 leaves, 16 cm. (8vo)

**Aristotle**

Historia de animalibus, Iulio Caesare Scaligero interprete, cum eiusdem commentariis.

Tolosae: Apud Dominicum & Petrum Bosc., 1619, [31], 1248, [22] pp., illus., 36 cm. (fol.)


**Aristotle**

On the parts of animals, trans. by W. Ogle.

London: Kegan Paul, 1882, xxxv, 263 pp., 26 cm.


**Aristotle**

History of animals. In ten books. Trans. by Richard Cresswell.

London: Bell, 1897, ix, 326 pp., 19 cm.


**Aristotle**

The works of Aristotle, trans. into English under the editorship of J. A. Smith and W. D. Ross.

Oxford: Clarendon P., 1910 – 12, vols. 4 & 5, 23 cm.


**Bacon,** Francis, *Viscount St. Albans* [1561 – 1626]

New Atlantis: a work unfinished.

London: Printed for William Lee, 1670, 31 pp., 29 cm. (4to)

[pp. 17 – 20 missing]


**Bacon,** Francis, *Viscount St. Albans*

History natural and experimental of life & death: or, of the prolongation of life.

London: Printed for William Lee, 1669, [7], 64 pp., 29 cm. (4to)


**Bacon,** Francis, *Viscount St. Albans*

Sylva sylvarum, or, a natural history, in ten centuries, whereunto is newly added . . . Articles of Enquiry touching metals and minerals.

London: Published after the author's death by William Rawles, 1670, [16], 14, 228 pp., 29 cm. (4to)

[Loose frontispiece and title page dated 1669. The index is separated from the main text by one of the other works by Bacon bound up in this copy]

## Barham, Henry [? – 1726]

An essay upon the silk-worm: containing many improvements upon this curious subject.

London: Printed for J. Bettenham & T. Bickerton, 1719, [12], 180, [7] pp., 18 cm. (8vo)


## Bartholomew Anglicus

Medieval lore: an epitome of the science, geography, animal and plant folk-lore and myth of the middle age, ed. by Robert Steele.

London: Elliot Stock, 1893, x, 154 pp., 24 cm.


## Belon, Pierre [1517 – 64]

L'histoire de la nature des oyseaux, avec leurs descriptions et naifs portraicts retirez du naturel escrite en sept livres.

Paris: Guillaume Cavellat, 1555, [22], 376 pp., illus. (col.), 32 cm. (fol.)

[Missing title page, dedication, pp. 159 – 74, 201 – 4, 223 – 6, 359 – 66, 377 – 81]


## Belon, Pierre

La nature et diversité des poissons.

Paris: Charles Estienne, 1555, [37], pp. 3 – 6, 11 – 14, 19 – 448, illus., 11 × 16 cm. (8vo)


## Belon, Pierre

Les observations de plusiers singularitez & choses memorables, trouvées en Grece, Asie, Indée, Egypte, Arabie, & autres pays estranges, redigées en trois livres.

Paris: Chez Guillaume Cavellat, 1555, [12], 211, [1] leaves, illus., 1 pl. (folded), 22 cm. (4to)


## Blankaart, Steven [1650 – 1702]

Schou-burg der Rupsen, Wormen, Ma'den en Vliegende Dierkens, daar uit voortkomende.

Amsterdam: Jan Ten Hoorn, 1688; [6], 234, [2] pp., front, 22 pls. (3 partly coloured), 16 cm. (8vo)

The heron from Belon's *L'histoire de la nature des oyseaux,* Paris, 1555

## **Bock,** Hieronymus [1498 – 1554]

De stirpium, maxime earum, quae in Germania nostra nascuntur.

[Strasbourg: s.n.], 1552, pp. 11 – 27, 1 – 1160, illus. (col.), 23 cm. (8vo)

## **Bodenheimer,** F. S.

Materialien zur Geschichte der Entomologie bis Linné. Band 1.

Berlin: Junk, 1928, x, 498 pp., illus., 24 pls., 29 cm.

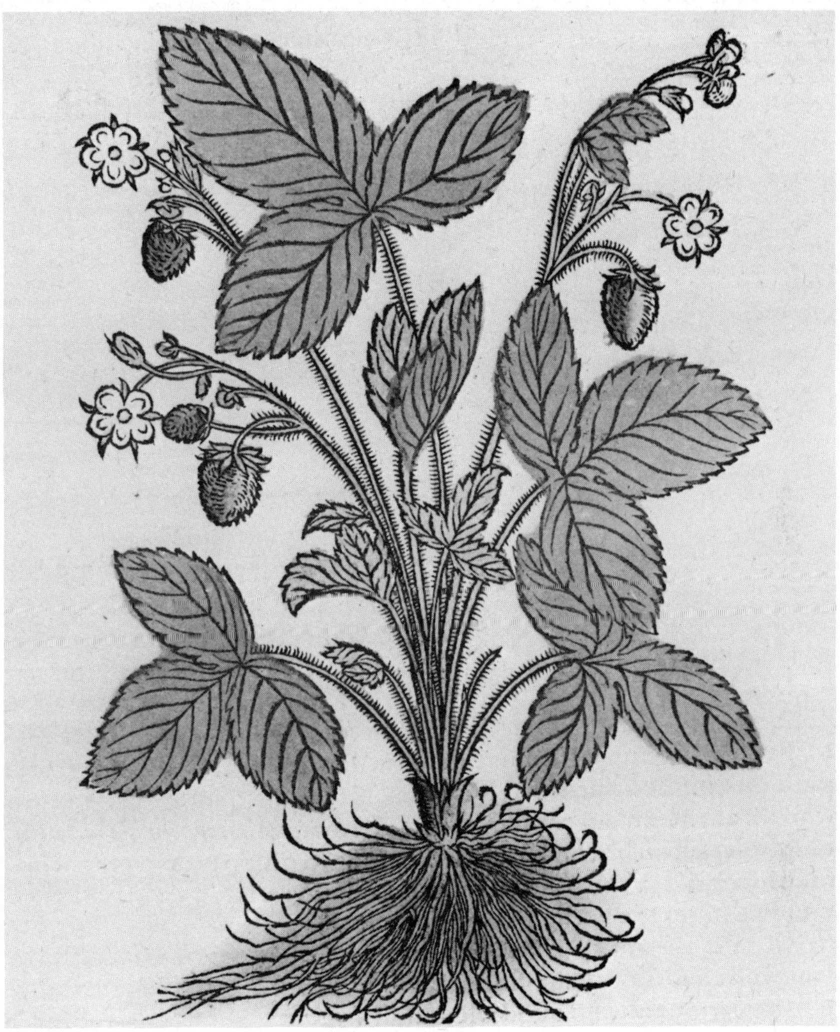

The strawberry plant in Bock's *De stirpium,* Strasbourg, 1552

## Brightwell, *Miss*

A life of Linnaeus.

London: van Voorst, 1858, vii, 191 pp., 16 cm.

**Brunfels** Otto [c. 1489 – 1534]

Herbarum vivae eicones, tomus II.

Argent: apud Ionnem Schottum, 1532, [2], 90, 199, [5] pp., illus., 32 cm. (fol.)


**Bruyerino,** Ioannes

De re cibaria libris XXII: omnium ciborum genera, omnium gentium moribus, & usu probata complectentes. Prima editio.

Lugduni: Apud Sebast. Honoratum, 1560, [24], 1130 pp., 17 cm. (8vo)


**Butler,** Charles [1571 – 1647]

The feminine monarchie or a treatise concerning bees, and the due ordering of them.

Oxford: printed by Joseph Barnes, 1609, [238] pp., illus., 15 cm. (8vo)


**Butler,** Charles

Thè feminine monarchie, or a treatise concerning bees. [2nd edition].

London: [s.n.], 1623, [12], 160 pp., illus., 20 cm. (4to)

[Imperfect copy, lacking original title page and pages after p. 160]


**Caius,** John [1510 – 73]

De canibus Britannicis, liber unus, [etc].

London: Car. Davis, 1729, xv, 249, [7] pp., 18 cm. (8vo)


**Caius,** John

Of English dogges, the diversities, the names, the natures, and the properties.

[s.l.: s.n.], 1881 reprint of 1576 edition, [10], 44, [7] pp., 21 cm.


**Camerarius,** Joachimus [1534 – 98]

Hortus medicus et philosophicus: in quo plurimarum stirpium breves descriptiones, . . .

Francofurti ad Moenum: Apud Hannem Feyerabend, 1588, [15], 184 pp., 20 cm. (4to)

**Cantimpratensis,** Thomas (Thomas of Cantimpré)
[fl. 13th cent.]

Liber apum.

[s.l.: s.n.],? c. 1500, [6], 94 leaves, 24 cm. (8vo)


**Cardano,** Girolamo (Hieronymus Cardanus) [1501 – 76]

De subtilitate libri XXI.

Lugduni: Apud Gulielmum Rouillium, 1559, 718, [54] pp., illus., 17 cm. (8vo)


**Cardano,** Girolamo

Les livres de Hierome Cardanus . . ., traduits de Latin en Françoys, par Richard le Blanc.

Paris: Gilles Beys, 1578, [36], 478 leaves, illus., 17 cm. (8vo)


**Cardano,** Girolamo

*See also*
Scaligeri, Julius Caesar


**Cardanus,** Hieronymus

*See*
Cardano, Girolamo


**Carus,** Titus Lucretius [fl. 1st cent. B.C.]

Of the nature of things. A metrical translation by W. E. Leonard.

London: Dent, 1916, xv, 301 pp., front., 21 cm.


**Cato,** Marcus Porcius [234 – 149 B.C.]

Cato's farm management . . . with notes . . . by a Virginia farmer.

Privately printed, 1910, 60 pp., 20 cm.


## Certain Ancient Tracts

Certain ancient tracts concerning the management of landed property reprinted.

London: Printed for C. Bathurst & J. Newbery, 1767, [324] pp. [var. pag.], 22 cm. (8vo)

**Clutius, Angerius**

*See*
Cluyt, Dick Outger


**Clutius, Theodore**

*See*
Cluyt, Theodore


**Cluyt, Dick Outger (Angerius Clutius)**

Opuscula duo singularia. I. De nuce medica. II. de hemerobio sive ephemero insecto, & majali verme.

Amsterodami: Typis Jacobi Charpentier, 1634, [38], 106 pp., illus., 1 pl. (folded), 20 cm. (4to)


**Cluyt, Theodore (Theodore Clutius)**

Van de Biën, Haeren wonderlycken oorspronck, natuer, eygenschap, krachtige ongehoorde ende selsaeme Wercken.

Antwerp: Andreas Paulas Colpyn, ?1618, [8], 205, [11] pp., 15 cm. (8vo)


**Columella, Lucius Junius Moderatus [fl. 1st cent. A.D.]**

De re rustica libri XII.

Paris: Robert Stephan, 1543, 498, [20] pp., illus., 18 cm. (8vo)


**Columella, Lucius Junius Moderatus**

Of husbandry. In twelve books: and his book concerning trees.

London: Printed for A. MIllar, 1745, xiv, [14], 600, [8] pp., 27 cm. (4to)


**Crescentiis, Petrus de [1230 – 1310]**

Opera d'agricoltura.

Venegia: Bernadino de Viano, 1538, [782] pp., illus., 16 cm. (8vo)


**Curtis, John [1791 – 1862]**

British entomology; being illustrations and descriptions of the genera of insects found in Great Britain and Ireland, vol. 5. Lepidoptera, part 1.

London: Printed for the author, 1823 – 40 [200] pp., 93 pls. (col.), 25 cm. (4to)

## Curtis, William [1746 – 99]

Instructions for collecting and preserving insects; particularly moths and butterflies.

London: Printed for the author, 1771, iv, 44 pp., 1 pl., 20 cm. (4to)


## Curtis, William

Lectures on botany, as delivered in the Botanic Garden at Lambeth.

London: Printed for H. D. Symonds and Curtis, 1805, vols. 1 & 3, 23 cm. (8vo)


## Curtis, William

Fundamenta entomologiae.

*See*
Linnaeus, Carl


## Denham, Joshua Frederick

A memoir of Francis Willughby, Esq. F.R.S. the naturalist

[s.l.]: Naturalist's Library, 1838, [4], pp. 17 – 146, front., 17 cm.


## Derham, W.

*See*
Albin, Eleazar


## Dodoens, Rembert [1516 – 85]

A niewe herball, or historie of plantes . . . And nowe first translated out of French into English, by Henry Lyte Esquyer.

London: Gerard Dewes, 1578, [22], 779, [17] pp., illus., 31 cm. (fol.)

[Lacks title page]


## Drury, Dru [1725 – 1803]

Illustrations of natural history. Wherein are exhibited upwards of two hundred and forty figures of exotic insects, according to their different genera.

London: Printed for the author, 1770, xviii, 130 pp., + 3pp. of handwritten index, front., 50 pls. (col.), 31 cm. (fol.)

**Edwardes,** Tickner

The lore of the honey-bee.

London: Methuen, 1908, xxiv, 281 pp., front., 23 pls., 20 cm.


**Edwardes,** Tickner

The lore of the honey-bee. 8th edition.

London: Methuen, 1917, xix, 196 pp., 17 cm.


**Estienne,** Charles [1504 – 64] *and* Jean **Liebault** [c. 1535 – 96]

L'agriculture ̈et maison rustique.

Luneville: Par Charles de la Fontaine, 1578, [8], 459, [27] pp., illus., 25 cm. (fol.)


**Estienne,** Charles *and* Jean **Liebault**

Maison rustique, or the countrie farme. Translated into English by Richard Surflet.

London: Printed by Edm. Bollifant, 1600, [26], 901, [26] pp., illus., 23 cm. (8vo)


**Fenn,** *Lady* Eleanor [1743 – 1813]

A short history of insects, (extracted from works of credit) designed as an introduction . . . Leverian Museum.

Norwich: Stevenson & Matchett, [1797], xxiv, 107 pp., front. (col.), 6 pls. (col.), 19 cm. (12 mo)


**Flower,** *Sir* William Henry

Essays on museums and other subjects connected with natural history.

London: Macmillan, 1898, xv, 394 pp., illus., 23 cm.


**Frampton,** John

*See*
Monardes, Nicolas

The cowslip in Fuch's *Plantarum effigies,* Lyon, 1552

**Freigius,** Joannes Thomas

Quaestiones physicae. In quibus methodus doctrinam physicam legitime docendi, describendique rudi Minerva descripta est, libris XXXVI.

Basileae: Sebastianum Henricpetri, 1585, 1295 pp., illus., 17 cm. (8vo)


**Fries,** Johann Jacob [1547 – 1611]

*See*
Gesner, Conrad

Bibliotheca instituta . . .


**Fuchs,** Leonhard [1501 – 66]

De historia stirpium commentarii insignes, maximis impensis et vigiliis elaborati.

Basileae: in officina Isingriniana, 1542, [26], pp. 12 – 888, lacking pages in between, illus. (col.), 37 cm. (fol.)


**Fuchs,** Leonhard

Plantarum effigies.

Lyon: Balthazar Arnoullet, 1552, [10], 516 pp., illus., 12 cm. (8vo)

[Lacks title page, pp. 111 – 2, 127 – 8 & part of index; spine reads "Erbario portatile"]


**Gardner,** Willoughby [1860 – 1953]

A list of the Hymenoptera – Aculeata so far observed in the counties of Lancashire and Cheshire, with notes on the habits of the genera.

(Repr. from "Trans. Liverpool Biol. Soc.", vol. 15, 1901, 62 pp., 1 map)


**Gedde,** John

A new discovery of an excellent method of bee-houses and colonies.

London: printed by D. Newman, 1675, [8], 30 pp., 1 fig. (folded), 16 cm. (8vo)


**Gedde,** John

The English apiary: or, the compleat bee-master . . . with a new discovery of an excellent method for making bee-houses and colonies.

London: Printed for E. Curll and others, 1721, [24], 108 pp., front., 17 cm. (8vo)

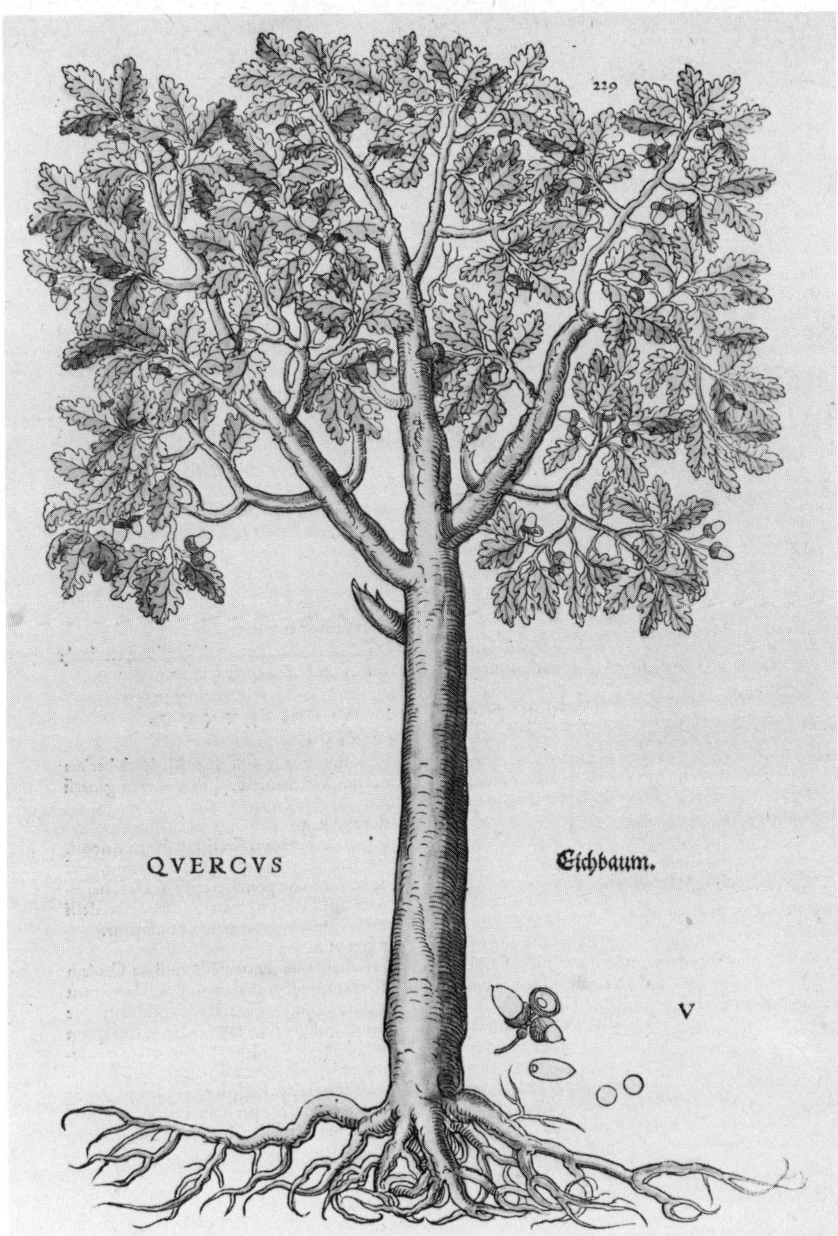


Fuch's illustration of the oak in his *De historia stirpium*, Basle, 1542

## Geoponika

Geoponika: agricultural persuits, trans. from the Greek by the Rev. T. Owen:

London: Printed for the author, 1805 – 6, 2 v., 22 cm.


## Gerard, John [1545 – 1612]

The herball or generall historie of plantes . . . very much enlarged and amended by Thomas Johnson.

London: Printed by Adam Islip, Joice Norton and Richard Whitakers, 1633, [38], 1630, [48] pp., illus., 35 cm. (fol.)

[Missing pp. 1599 – 1600; spine reads "Johnson's Herbal"]


## Gesner, Conrad [1516 – 65]

Historia plantarum et vires ex Dioscoride, Paulo Aegineta, Theophrasto, Plinio, & recentioribus graecis, iuxta elementorum ordinem.

Basileae: Apud Robertum Wynter, 1541, [7], 281, [15] pp., 15 cm. (8vo)

[Stamped binding dated 1553]


## Gesner, Conrad

Bibliotheca universalis, sive catalogus omnium scriptorum locupletissimus, in tribus linguis, Latina, Graeca & Hebraica.

Tiguri: Apud Christophorum Froschoverum, 1545, [18], 631 leaves, 34 cm. (fol.)


## Gesner, Conrad

Appendix bibliothecae Conradi Gesneri.

Tiguri: apud Christophorum Froschoverum, 1555, [8], 105, [1] leaves, 31 cm. (fol.)


## Gesner, Conrad

Bibliotheca instituta et collecta, primum a Conrado Gesnero . . . amplificata, per Iohannem Iacobum Frisium.

Tiguri: Excudebat Christophorus Froschoverus, 1583, [56], 835, [3] pp., 34 cm. (fol.)

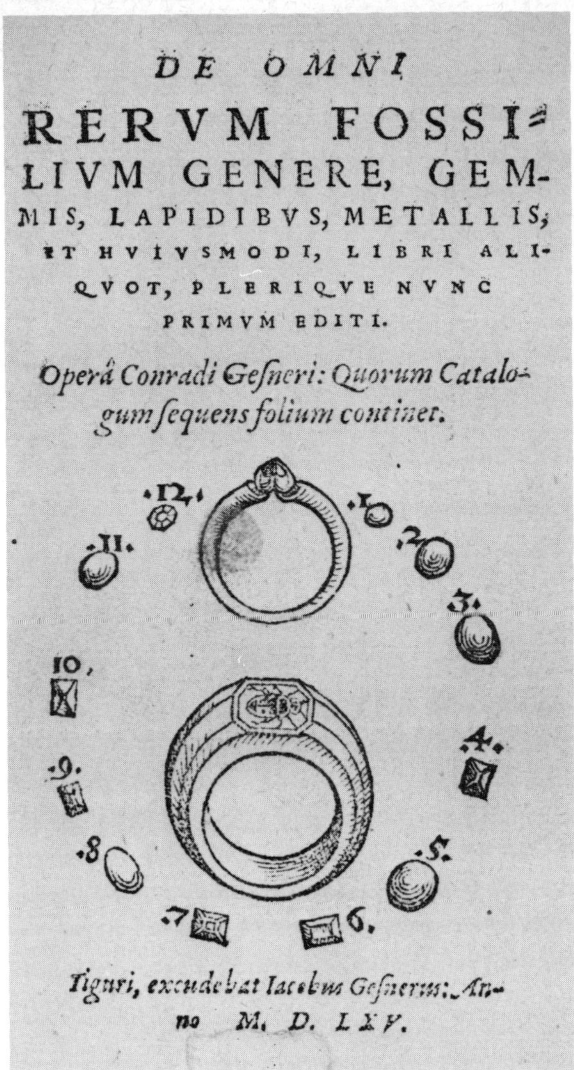



The title page to Gesner's *De omni rerum fossilium*, Zurich, 1565

## Gesner, Conrad

Historiae animalium lib. I. de quadrupedibus viviparis.

Tiguri: Apud Christ. Froschoverum, 1551, [40], 1104, [10] pp., illus. (col.), 37 cm. (fol.)

**Gesner,** Conrad

Historiae animalium liber primus. De quadrupedibus viviparis. Editio secunda.

Francofurti: In bibliopolio Cambieriano, 1602, [40], 967 pp., illus., 36 cm. (fol.)


**Gesner,** Conrad

Historiae animalium liber II. de quadrupedibus oviparis.

Tiguri: Excudebat C. Froschoverus, 1554, [6], 110, 27 pp., illus. (col.), 37 cm. (fol.)

[The illustrations of the copy bound with "Historiae animalium liber III" are not coloured]


**Gesner,** Conrad

Historiae animalium liber II. qui est de quadrupedibus oviparis.

Francofurdi: Ex officina typographica Ionnis Wecheli, 1586, [6], 119 pp., illus., 37 cm. (fol.)


**Gesner,** Conrad

Historiae animalium liber III. qui est de avium natura.

Tiguri: Apud Christoph. Froschoverum, 1555, [34], 779 pp., illus. (some col.), 37 cm. (fol.)


**Gesner,** Conrad

Historiae animalium liber III. qui est de avium natura.

Francofurdi: Ex officina typographica Ionnis Wecheli, 1585, [12], 806, [26] pp., illus., 37 cm. (fol.)


**Gesner,** Conrad

Historiae animalium liber IIII. qui est de piscium & aquatilium animantium natura.

Tiguri: Apud Christoph. Froschoverum, 1558, [40], 1297 pp., illus. (col.), 37 cm. (fol.)

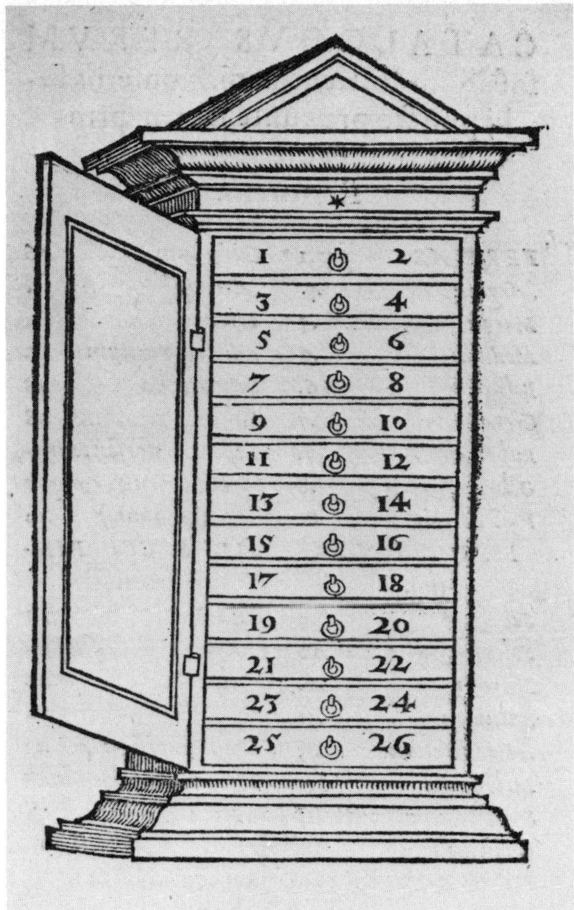


Kentmann's museum cabinet in Gesner's
*De omni rerum fossilium,* Zurich, 1565

## **Gesner,** Conrad

Historiae animalium liber IV. Qui est de piscium & aquatilium animantium natura.

Francofurti: In bibliopolio Andreae Cambieri, 1604, [40], 1052, 38 pp., illus., 36 cm. (fol.)


## **Gesner,** Conrad

Historiae animalium lib. V. qui est de serpentium natura.

Tiguri: In officina Froschoviana, 1587, [6], 64, 11 leaves, illus., 37 cm. (fol.)

**Gesner,** Conrad

De piscibus et aquatilibus omnibus libelli III novi.

Tiguri: Apud Andream Gesnerum, 1556, [6], 280 pp., 16 cm. (8vo)


**Gesner,** Conrad

Icones animalium quadrupedum viviparorum et oviparorum . . . Editio secunda.

Tiguri: Excudebat C. Froschoverus, 1560, 127, [7] pp., illus. 39 cm. (fol.)


**Gesner,** Conrad

Icones avium omnium, quae in historia avium Conradi Gesneri describuntur. Editio secunda.

Tiguri: Excudebat C. Froschoverus, 1560, 137, [9] pp., illus., 39 cm. (fol.)


**Gesner,** Conrad

Nomenclator aquatilium animantium. Icones animalium aquatilium in mari & dulcibus aquis degentium . . .

Tiguri: Excudebat Christoph. Froschoverus, 1560, [28], 374, [1] pp., illus., 39 cm. (fol.)


**Gesner,** Conrad

De omni rerum fossilium genere, gemmis, lapidibus, metallis, et huiusmodi, libri aliquot, plerique nunc primum editi.

Tiguri: Excudebat Iacobus Gesnerus, 1565, [7], 169, [8], 95, [2], 22 leaves, illus., 18 cm. (8vo).

[A compendium of works, but this copy has only the two by Johann Kentmann, together with Gesner's main work]


**Gesner,** Conrad

*See also*
Simmler, Josias


**Goedaert,** Johann [c. 1617 – 68]

Metamorphosis et historia naturalis insectorum.

Medioburgi: Apud Jacobum Fierensium, 1662 – 67, 3 v., front., 126 pls., 16 cm. (8vo)

**Goedaert, Johann**

Metamorphosis et historia naturalis insectorum.

Medioburgi: Apud Jacobum Fierensium, 1662 – 67, v. 1 & 2, front., (col.), 106 pls. (col.), 16 cm. (8vo)

[Coloured heraldic device of Henri Thibaut on p. xxiv of vol. 1]


**Goedaert, Johann**

Metamorphosis naturalis ofte Historische Beschryvinge van den Oirspronk.

Middelburgh: Gedruckt Jaques Fierens, 1669, 3 v., front. (col.), 126 pls. (col.), 16 cm. (8vo)


**Goedaert, Johann**

Metamorphoses naturelles ou historie des insectes.

La Haye: Chez Adrian Moetiens, 1700, 3 v., front., 127 pls., 16 cm. (8vo)


**Goedaert, Johann**

Metamorphoses naturelles ou histoire des insectes.

Amsterdam: Chez Pierre Mortier, 1700, 3 v., fronts., 127 pls., 15 cm. (8vo)


**Goedaert, Johann**

Metamorphosis et historia naturalis insectorum.

Medioburgi: Apud Jacobum Fierensium, ?1700, 3 v., front., 126 pls., (col.), 16 cm. (8vo)


**Goedaert, Johann**

De insectis, in methodum redactus; cum notularum additione. [Appendix by Martin Lister].

London: Excudebat R. E. sumptibus S. Smith, 1685, [8], 356, [4], 45 pp., 21 pls., 20 cm. (8vo)


**Goedaert, Johann**

Of insects.

York: Printed by John White, 1682, [6], 140 pp., 14 pls., 21 cm. (4to)

**Googe,** Barnaby

The whole art and trade of husbandry, contained in foure bookes.

London: Printed by T. S. for Richard More, 1614, [11], 183 leaves, 18 cm. (8vo)


**Gould,** William

An account of English ants.

London: Printed for A. Millar, 1747, [16], 109 pp., 16 cm. (8vo)


**Gt. Britain & Ireland.** Historical Manuscripts Commission

Report on the manuscripts of Lord Middleton, preserved at Wollaton Hall, Nottinghamshire.

London: H.M.S.O., 1911,xv, 746 pp., 24 cm.

[pp. 269 – 72, 503 – 610 missing]


**Gunther,** R. T.

Early British botanists and their gardens.

Oxford: The author, 1922, viii, 417 pp., front., illus., 8 pls., 26 cm.


**Harris,** Moses [c. 1731 – c. 1785]

The English Lepidoptera: or, the aurelian's pocket companion.

London: Printed for J. Robson, 1775, xvi, 66 pp., front (col.), 22 cm. (4to)


**Harris,** Moses

The Aurelian. A natural history of English moths and butterflies, together with the plants on which they feed.

London: Printed for the author, 1766 [?1794 issue], [2], 145, [4] pp., 45 pls. (col.), 44 cm. (fol.)


**Harris,** Moses

An exposition of English insects, with curious observations and remarks.

London: Printed for the author, ?1782, [4], 166 [4] pp., fronts. (col.), 50 pls. (col.) + 1 pl., 29 cm. (fol.)

**Hartlib,** Samuel [? d.1670]

His legacie: or an enlargement of the discourse of husbandry used in Brabant & Flaunders. 2nd edition.

London: Printed by R. & W. Leybourn, 1652, [8], 118, [28] pp., 19 cm. (4to)

**Hartlib,** Samuel

The reformed common-wealth of bees. Presented in severall letters and observations to Samuel Hartlib Esq.

London: Printed for Giles Calvert, 1655, [4] 62, [2] pp., illus., 18 cm. (4to)

**Herodotus** [c. 484 – 425 B.C.]

Herodotus: a new and literal version by Henry Cars.

London: Bohn, 1848, vi, 613 pp., front., 19 cm.

**Hesiod** [fl. c. 800 B.C.]

The Homeric hymns and Homerica, trans. by Hugh G. Evelyn-White.

London: Heinemann, 1920, xlviii, 639 pp., front., 17 cm.

**Hieronymus,** Marcus

Opera.

Venetiis: per Melchiorem Sessam, 1538, 335, [29] pp., 16 cm. (8vo)

**Hieronymus,** Marcus

Opera.

Lugduni: Apud Antonium Gryphium, 1586, 573 pp., 13 cm. (8vo)

**Hill,** Thomas

A pleasaunt instruction of the parfit ordering of bees.

London: Imprinted by Thomas Marshe, 1568, [6], 54 leaves, 15 cm. (8vo)

## Hill, Thomas

A pleasaunt instruction of the parfit orderinge of bees.

[s.l.: s.n.], 1568, [6], 64 leaves, front., 13 cm. (8vo)

[This copy is different from the other 1568 edition in style of pagination, front., and lay-out of title page]


## Hill, Thomas

A profitable instruction of the perfite ordering of bees.

London: Imprinted by Henrie Bynneman, 1579, [102] pp., 20 cm. (4to)


## Hill, Thomas

The proffitable arte of gardening.

London: Imprinted by Thomas Marshe, 1568, 194 leaves, illus., 1 pl., 15 cm. (8vo)

[Lacks original title page, signatures A, B, fol. 15]


## Hill, Thomas

The profitable arte of gardening.

London: Imprinted by Henrie Bynneman, 1579, 152 pp., illus., 20 cm. (4to)

[Lacks original title page, signatures A, B]


## Hoefnagel, D. I.

Diversae insectarum volatilium icones ad vivum accuratissime depictae per celeberrimum pictorem.

[Amsterdam]: Nicolao Ioannis Visscher, 1630, title page, inserted frontispiece depicting Hoefnagel, 15 pls., 21 × 30 cm. (4to)


## Hoefnagel, George [1545 – ?1617]

Archetypa studiaque Patris Georgii Hoefnagelii Iacobus F: genio duce ab ipso scalpta, omnibus philomusis amicé D: ac Perbenigé communicat. Ann. Sal. XCII. Aetat: XXII.

Francofurti ad Maenum: [s.n.], 1592, title page, 51 pls., 20 × 27 cm. (4to)

## Hollar, Wenceslaus [1607 – 77]

Muscarum scarebeorum vermiumque varie figure et formae omnes primo ad vivum coloribus depictae et ex collectione Arundeliana.

Antwerp: [s.n.], 1646, title page and 12 engraved plates, remounted, 14 × 20 cm. (8vo)

## Hortus Sanitatis

Ortus sanitatis. De herbis et plantis. De animalibus et reptilibus. De avibus et volatilibus. De piscibus et natatilibus. De lapidibus et in terre venis nascentibus. Urinis et earum speciebus.

[s.l.: s.n.], 1517, [349] leaves, illus. (col.), 32 cm. (fol.)

## Huber, François [1750 – 1831]

New observations on the natural history of bees.

Edinburgh: Printed for John Anderson, 1806, 300 pp., 19 cm.

## Imperato, Ferrante [1550 – 1625]

Dell' historia naturale di Ferrante Imperato Napolitano. Libri XXVIII

Napoli: Per Constantino Vitale, 1599, [28], 791 pp., illus., 30 cm. (fol.)

## Isidore *of Seville* [c. 560 – 636]

De natura rerum, ed. by Gustavus Becker.

Berolini: Weidmann, 1857, xxxii, 88 pp., 1 fig., 25 cm.

## Jonson, Thomas [c. 1600 – 44]

Mercurius botanicus. Sive plantarum gratia suscepti itineris, ed. by T. S. Ralph.

Londini: Guliel. Pamplin, 1849, [6], 78, [7], 37 pp., 22 cm.

## Jonston, John [1603 – 75]

Thaumatographia naturalis, in decem classes distincta.

Amsterdami: Guilielmum Blaeu, 1632, [12], 501, [3] pp., 14 cm. (12mo)

## Jonston, John

Thaumatographia naturalis, in classes decem divisa. Editio secunda priore auctior.

Amstelodami: Apud Ioannem Ianssonium, 1633, [6], 578, [2] pp., 13 cm. (12mo)


## Jonston, John

Historiae naturalis de insectis libri III, [with] De serpentibus et draconibus libri II.

Amstelodami: Apud Ioannem Iacobi Fil. Schipper, 1657, [8], 148, [2], 38 pp., 26 pls., 37 cm. (fol.)


## Jonston, John

Historiae naturalis, de quadrupedibus, [with] De insectis libri III, de serpentibus et draconibus libri II.

Amstelodami: Apud Joannem Jacobi Fil. Schipper, 1657, 6, [2], 164, [8], 148, [2], 38 pp., 116 pls., 39 cm. (fol.)


## Jonston, John

An history of the wonderful things of nature set forth in ten severall classes.

London: Printed by John Streater, 1657, [16], 354 pp., 28 cm. (4to)


## Jonston, John

Theatrum universale omnium insectorum, [with] Historia naturalis de serpentibus libri duo.

Heilbronnae: Franciscus Iosephus Eckebrecht, 1757, [8], 212, [4], 55, [2] pp., 40 pls., 34 cm. (fol.)


## Kentmann, Johann [1518 – 74]

Catalogus rerum fossilium Io. Kentmani
*and*
Calculorum qui in corpore ac membris hominum innascuntur, genera XII.
*See*
Gesner, Conrad
De omni rerum fossilium.

## Keys, John

The antient bee-masters farewell; or, full and plain directions for the management of bees to the greatest advantage.

London: Printed for G. G. and J. Robinson, 1796, xvi, 273 pp., front., 1 pl., 13 cm. (8vo)


## Kirby, William [1759 – 1850]

Monographia apum Angliae.

Ipswich: Printed for the author, 1802, 2 v., 4 pls. (col.) + 14 pls., 22 cm.


## Kirby, William *and* William Spence [1783 – 1860]

An introduction to entomology; or, elements of the natural history of insects. 7th edition.

London: Longman, 1856, xxviii, 607 pp., 19 cm.


## Kirby, William *and* William Spence

An introduction to entomology; or, elements of the natural history of insects. 7th edition.

London: Longman, 1857, xxviii, 607 pp., 19 cm.


## Klebs, Arnold C.

A catalogue of early herbals, mostly from the well-known library of Dr. Karl Becher, Karlsbad.

Zurich: Art Ancien, 1925, xxiv, 61 pp., front., illus., 2 pls., 25 cm.


## Kreuter Buch

[An early German herbal].

[s.l.: s.n., c. 1520], [149] leaves, illus. (col.), 31 cm. (fol.)

[Begins at signature C; leaf XI torn away]


## L'Admiral, Jacob [1700 – 70]

Naauwkeurige waarneemingen omtrent de veranderingen van veele Insekten of gekorvene Diertjes.

Amsterdam: Johannes Sluyter, 1774, [4], 34, [2] pp., 33 pls. (col.), 43 cm. (fol.)

Part of the title page to Lawson's *A new orchard and garden,* London, 1631

## Lawson, William

A new orchard and garden or the best way for planting, grafting, and to make any ground good, for a rich orchard.

London: Printed by Nicholas Okes, 1631, [8], 134 pp., illus., 20 cm. (8vo)


## Lawson, William

A new orchard and garden or the best way for planting, grafting, and to make any ground good, for a rich orchard.

London: Printed by William Wilson, 1660 [6], 56 pp., illus. 19 cm. (4to)


## Lawson, William

A new orchard and garden.

London: Cresset Press, 1927, xxvi, 116 pp., illus., 22 cm.


## Lawson, William

The country house-wives garden, containing rules for herbs, and seeds, of common use.

London: Printed by William Wilson, 1660, [2], pp. 69 – 92, illus., 19 cm. (4to)

**Levett, John**

The ordering of bees: or, the true history of managing them from time to time.

London: Printed by Thomas Harper, 1634, [22], 71 pp., 18 cm. (4to)


**Lewin, W.** [? – 1795]

The insects of Great Britain, systematically arranged, accurately engraved, and painted from nature. Vol. 1.

London: Printed for J. Johnson, 1795, 97, [3] pp., 46 pls. (col.), 28 cm. (4to)


**Libri de re Rustica**

Libri de re rustica.

Venetiis: In Aedibus Aldi, 1514, [33], 308 leaves, illus., 22 cm. (8vo)


**Linnaeus, Carl** [1707 – 78]

Fundamenta entomologiae: or, an introduction to the knowledge of insects. Being a translation of the Fundamenta entomologiae of Linnaeus . . . by W. Curtis.

London: G. Pearch, 1772, viii, 90 pp., 2 pls., 21 cm. (4to)


**Linnaeus, Carl**

Systema naturae. Regnum animale. 10th edition, 1758.

Lipsiae: Guilielmi Englemann, 1894, [6], 824, iii pp., 22 cm.


**Lister, Martin** [1639 – 1712]

Historiae animalium Angliae tres tractatus.

London: Apud Joh. Martyn, 1678, [6], 250 pp., 9 pls., 19 cm. (4to)


**London.** Ray Society

Memorials of John Ray, ed. by Edwin Lankester.

London: Ray Society, 1846, xii, 220 pp., front., 23 cm.


**London.** Ray Society

The correspondence of John Ray, ed. by Edwin Lankester.

London: Ray Society, 1848, xvi, 502 pp., front., 23 cm.

**London.** Ray Society

Further correspondence of John Ray, ed. by R. W. T. Gunther.

London: Ray Society, 1928, xxiv, 332 pp., front., illus., 2 pls., 22 cm.


**Lyly,** John

Euphues. The anatomy of wit. Euphues and his England, ed. by E. Arber.

Birmingham: [s.n.]. 1868, 479 pp., 17 cm.


**Magnus,** Olaus [1490 – 1557]

Historiae septentrionalium gentium breviarum. Libri XXII. Editio nova.

Lugduni Batavorum: Apud Adrianum Wyngaerde et Franciscum Moiardum, 1645, [16], 589, [67] pp., 13 cm. (12mo)


**Malpighi,** Marcello [1628 – 94]

Dissertatio epistolica de bombyce.

London: Apud Joannem Martyn & Jacobum Allestry, 1669, [8], 100 pp., 12 pls., 24 cm. (4to)


**Markham,** Gervase [?1568 – 1637]

Cheape and good husbandry for the well-ordering of all beast and fowles. 10th edition.

London: Printed by W. Wilson, 1660, [8], 146, [10] pp., 19 cm. (4to)


**Markham,** Gervase

The generall cure and order of all horses, [etc.]

[s.l.: s.n.], n.d., 188 pp., 20 cm. (8vo)

[From an edition of 'Cheape and good husbandry']


**Martin,** Matthew [1748 – 1838]

The aurelian's vade mecum.

Exeter: Printed by R. Trewman, 1785, xii, [36] pp., 18 cm. (8vo)

**Mattioli,** Pietro Andrea [1501 – 77]

Commentarii in sex libros Pedacii Dioscoridis Anazarbei de Medica materia.

Venetiis: Ex officina Valgrisiana, 1565, [170], 1460, [12] pp., front., illus., 37 cm. (fol.)


**Mattioli,** Pietro Andrea

De plantis epitome utilissima.

Francofurti ad Moenum: [s.n.], 1586, [12] 1003, [27] pp., illus., 23 cm. (8vo)


**Maxwell,** Robert

The practical bee-master.

Edinburgh: Printed by Robert Drummond, 1747, viii, 138 pp., 17 cm. (4to)


**Maxwell Lyte,** H. C. *and* Edmund **Buckle**

The Lytes of Lytescary [and] Lytescary.

(Repr. from "Procs. Somersetshire Arch. & Nat. Hist. Soc.", vol. 18, 1892, pp. 1 – 110, illus., 4 pls., 3 figs.)


**Megenberg,** Conrad von

*See*
Alemann, Conrad von, *of Magdeburg*


**Merian,** Maria Sibylla [1647 – 1717]

Der Raupen wunderbare Verwandelung und sonderbare Blumen-nahrung.

Nurnberg: Johann Andreas Graffen, 1679 – 83, 2 v., 99 pls., 21 cm. (4to)

[Vol. 2 missing pp. 97 – 8 & pl. 50]


**Merian,** Maria Sybylla

Erucarum ortuns, alimentum et paradoxa metamorphosis.

Amsterdami: Apud Joanne Oosterwyk, [1717], [12], 64 pp., 150 pls., 25 cm. (4to)

## Merrett, Christopher [1614 – 95]

Pinax rerum naturalium Britannicarum, continens vegetabilia, animalia et fossilia.

London: Typis T. Roycroft, 1667, [32], 224 pp., 16 cm. (8vo)


## Merrett, Christopher

Pinax rerum naturalium Britannicarum, continens vegetabilia, animalia et fossilia. 2nd edition.

London: Typis T. Roycroft, 1667, [32], 224 pp., 16 cm. (8vo)


## Miall, L. C.

History of biology.

London: Watts, 1911, vii, 151 pp., front., illus., 19 cm.


## Mills, John

An essay on the management of bees.

London: Printed for J. Johnson and B. Davenport, 1766, xii, 157 pp., front., 21 cm. (8vo)


## Moberly, C. E.

The early Tudors. Henry VII: Henry VIII.

London: Longmans, 1887, xvi, 243 pp., front. (map), 15 cm.


## Moffet, Thomas [1553 – 1604]

Nosomantica Hippocratea.

Francofurdi: Apud heredes Andreae Wecheli, 1588, [15], 195 pp., 15 cm. (8vo)


## Moffet, Thomas

Insectorum sive minimorum animalium theatrum: olim ab Edoardo Wottono, Conrado Gesnero, Thomaque Pennio inchoatum.

Londini: ex officina typographica Thom. Cotes, 1634, [20], 326 pp., illus., 4 pls., 28, 30 and 31 cm. (fol.)

## Moffet, Thomas

The theater of insects: or, lesser living creatures.

London: Printed by E.C.,1658, [12], pp. 889 – 1150 + [6] pp., illus., 34 cm. (fol.)

[This work formed an appendix to Topsell's "History of four-footed beasts"]


## Moffet, Thomas

Healths improvement: or, rules comprizing and discovering the nature, method, and manner of preparing all sorts of food used in this nation. Corrected and enlarged by Christopher Bennet.

London: Printed by Tho: Newcomb, 1655, 8, 296 pp., 19 cm. (4to)


## Moffet, Thomas

Health's improvement: or, rules comprizing and discovering the nature, method, and manner of preparing all sorts of food used in this nation.

London: Printed for T. Osborne, 1746, xxxii, 398 pp., 17 cm. (12mo)

Beehive on the title page of Moffet's *Insectorum,*
London, 1634

## Monardes, Nicolas

Joyfull newes out of the newe founde worlde . . . englished by John Frampton.

London: Constable, 1925 reprint of 1577 edition, 2 v., 21 cm.


## Oppian, *of Anazarbus* [2nd Cent. A.D.]

De piscatu libri V. De venatione libri IIII.

Paris: Apud Adr. Turnebum, 1555, [8], 216, [2], 203 pp., 22 cm. (4to)

[pp. 149 – 56 missing in "De Venatione"]


## Oppian, *of Anazarbus*

De venatione lib. IIII. De piscatu lib. V. cum interpretatione . . . Conradi Rittershusii.

Lugduni Batavorum: Apud Franciscum Raphelengium, 1597, [88], 376, [32], 164, [3], [8], 344 pp., 16 cm. (8vo)


## Oppian, *of Anazarbus*

Oppian's halieuticks of the nature of fishes and fishing of the ancients in V. books.

Oxford: Printed at the Theater, 1722, [8], 13, [2], 232, [7] pp., 24 cm. (4to)

## Palladius

The fourteen books of Palladius Rutilius Taurus Aemilianus, on agriculture, by T. Owen.

London: Printed for J. White, 1807, [16], 349 pp., 21 cm.


## Palladius

On husbondrie. Ed. by the Rev. Barton Lodge.

London: Early English Text Society, 1873, [4], 220 pp., 23 cm.


## Persius (Aulus Persius Flaccus) [A.D. 34 – 62]

*See*
Stelluti, Francesco

## Petiver, James [1663 – 1718]

Musei Petiveriani centuria prima, rariora naturae continens: viz. animalia, fossilia, plantas.

Londini: Ex officina S. Smith & B. Walford, 1695 – 1703, 96, 16 pp., front., 1 pl., 19 cm. (8vo)

## Petiver, James

Hortus Peruvianus medicinalis: or, The south sea herbal.

[London: s.n., 1715], 3 pp., 6 pls., 38 cm. (fol.)

## Petiver, James

Plantarum Italiae. Plantarum Aegyptiacarum. Plantae Silesiacae. Plantarum Etruriae. Monspelli plantarum.

London: the author, 1715 – 17, [10] pp., 7 pls., 38 cm. (fol.)

## Petiver, James

Pteri-graphia Americana.

London: [s.n.] 1716, [3] pp., 20 pls., 38 cm. (fol.)

## Petiver, James

Papilionum Britanniae icones, nomina, etc. Containing the figures, names, places, seasons etc. of above eighty English butterflies.

London: the author, 1717, 2pp., 6 pls. (col.), 38 cm. (fol.)

## Petiver, James

Opera, historiam naturalem spectantia.

London: Printed for John Millan, 1767, 2v., [parts of, only], 18 pls. (col.) + 160 pls., 37, 38 cm. (fol.)

## Petiver, James

Herbarii Britannici . . .

*See*
Ray, John

## Pitfeild, Alexander

Memoirs for a natural history of animals. Containing the anatomical descriptions of several creatures dissected by the Royal Academy of Sciences at Paris.

London: Printed by Joseph Streater, 1688, [16] 267, [15], [2], 40 pp., front., 35 pls. 31 cm. (4to)

## Les Plaisirs Innocens

Les plaisirs innocens et amoureux de la campagne contenant le traité des mouches à miel, ou les régles pour les bien gouverner . . .

Amsterdam: Chez Paul Marret, 1699, 253, [7] pp., front., 1 pl., 15 cm. (12mo)

## Pliny (Caius Plinius Secundus) [A.D. 23 – 79]

Historia naturale di. C. Plino secundo tradocta i lingua Fiorentina per Christophoro Landino Fiorentino.

Venetiis: Opus Magistri Philippi, 1481, [287] leaves, 30 cm. (fol.)

## Pliny

Naturalis historiae.

Venetiis: Impressum Magistrum Marinus Saracenum, 1487, [265] leaves, 31 cm. (fol.)

## Pliny

De naturali historia.

Venetiis: Melchiore Sessam & Petrus Serenae, 1525, [305] leaves, illus., 30 cm. (fol.)

## Pliny

Naturalis historiae libros castigationes.

Basileae: Apud Ionnem Valderum, 1534, [83], 523 pp., 21 cm. (4to)

[Missing pp. 517 – 20]

## Pliny

Historiae naturalis libri XXXVII.

Lugduni Batavorum: Ex officina Elzeviriana, 1635, [24], 654, [18] pp., 13 cm. (12mo)

## Pliny

The natural history of Pliny, trans. by J. Bostock and H. T. Riley.

London: Bohn, 1855 – 7, 6 v., 19 cm.


## Power, Henry [1623 – 68]

Experimental philosophy, in three books.

London: Printed by T. Roycroft, 1664, [20], 193 pp., illus., 1 pl., 21 cm. (4to)


## Pullein, Samuel

*See*
Vida, Marcus Hieronymus


## Purchas, Samuel [c. 1577 – 1626]

A theatre of politicall flying-insects.

London: Printed by R. I., 1657, [22], pp. 1 – 214, 257 – 387. 19 cm. (4to)


## Raei, Joanne de

Clavis philsophiae naturalis, seu introductio ad naturae contemplationem, Aristotelico – Cartesiana.

Lugd. Batavor.: Ex officina Johannis & Danielis Elsevier, 1654, [28], 219 pp., illus., 19 cm. (4to)


## Ray, John [1627 – 1705]

Observations topographical, moral & physiological; made in a journey through part of the Low-Countries. . . .: with a catalogue of plants not native of England.

London: Printed for John Martyn, 1673, [15], 499, [7], 115 pp., 2 pls., 19 cm. (8vo)


## Ray, John

Methodus insectorum: seu insecta in methodum aliqualem digesta.

Londini: Apud Sam. Smith, 1705, [2], 16 pp., 19 cm. (4to)

**Ray,** John

Historia insectorum.

Londini: Impensis A. & J. Churchill, 1710, xv, 400 pp., 24 cm. (4to)


**Ray,** John

Herbarii Britannici clariss. D. Raij catalogus cum iconibus ad vivum delineatis
. . . a Jacobo Petiver.

[London: s.n., 1712], [11] pp., 72 pls., 38 cm. (fol.)


**Ray,** John

Philosophical letters between the late learned Mr. Ray and several of his
ingenious correspondents, native and foreigners.

London: W. Derham, 1718, [8], 376, [12] pp., illus., 20 cm. (4to)


**Ray,** John

*See also*
London. Ray Society


**Reaumur,** R. A. de [1683 – 1757]

Memoires pour servir à l'histoire des insectes.

Paris: de l'imprimerie royale, 1734 – 42, 6 v., 267 pls., 26cm. (4to)


**Redi,** Francesco [1626 – c. 1697]

Experimenta circa generationem insectorum.

Amstelodami: Sumptibus Andreae Frisii, 1671, [12], 330, [18] pp., illus.,
37 pls., 14 cm. (12mo)


**Redi,** Francesco

Experimenta circa res diversas naturales, speciatim illas, quae ex Indiis
adferuntur.

Amstelodami: Sumptibus Andreae Frisii, 1675, [2], 193, [13], 111, [7], 72pp.,
illus., 12 pls., 14 cm. (12mo)

**Redi,** Francesco

Esperienze intorno alla generazione degl' insetti fatte da Francesco Redi.

Napoli: Nella stamperia di Giacomo Raillard, 1687, [12], 195 pp., front., illus., 29 pls., 16 cm. (8vo)


**Redi,** Francesco

Experiments on the generation of insects. Trans. from the 1688 edition by Mab Bigelow.

Chicago: Open Court, 1909, 160 pp., front., illus., 22 cm.


**Remnant,** Richard

A discourse or historie of bees.

London: Printed by Robert Young, 1637, [4], 47 pp., 19 cm. (4to)


**Reymond,** Arnold

History of the sciences in Greco-Roman antiquity.

London: Methuen, 1927, x, 145 pp., illus., 20 cm.


**Roesel von Rosenhof,** August Johann [1705 – 59]

Der Monatlich herausgegebenen Insecten-Belustigung.

Nurnberg: Fleischmann, 1746 – 61. 4 v., 333 pls. (col.), 22 cm. (4to)


**Rohde,** Eleanour Sinclair

The old English herbals.

London: Longmans, 1922, xii, 243 pp., front. (col.), 17 pls., 26 cm.


**Rondelet,** Guillaume [1507 – 66]

Libri de piscibus marinis, in quibus verae piscium effigies expressae sunt.

Lugduni: Apud Matthiam Bonhomme, 1554, [14], 583, [23] pp., illus., 36 cm. (fol.)

## **Rondelet,** Guillaume

Universae aquatilium historiae pars altera, cum veris ipsorum imaginibus.

Lugduni: Apud Matthiam Bonhomme, 1555, [12], 242, [9] pp., illus., 36 cm. (fol.)


## **Rondelet,** Guillaume

L'histoire entiere des poissons. [2 parts].

Lion: Par Macé Bonhome, 1558, [10], 418, [14], [4], 181, [9] pp., illus., 29 cm. (fol.)


## **Roscoe,** William

The butterfly's ball and the grasshopper's feast. A facsimile reproduction of the edition of 1808.

London: Griffith & Farran, 1883, x, 11 pp., front., 6 pls. 22 cm.


## **Royds,** Thomas Fletcher

The beasts, birds and bees of Virgil: a naturalist's handbook to the Georgics. 2nd ed., revised.

Oxford: Blackwell, 1918, xix, 107 pp., 19 cm.


## **Rucellai,** M. G.

Le apidi.

Roma: [s.n.], 1541, 20 leaves, 15 cm. (4to)


## **Rusden,** Moses

A further discovery of bees.

London: Printed for the author, 1679, [24], 143 pp., front., 3 pls., 16, 17 cm. (8vo)


## **Salviani,** Ippolito [1514 – 72]

Aquatilium animalium historiae, liber primus, cum eorumdem formis, aere excusis.

Romae: Apud Hippolytum Salvianum, 1554 – 8, [8], 256 leaves, illus., 43 cm. (fol.)

[Colophon date 1558]

**Scaligeri,** Julius Caesar [1484 – 1558]

Exotericarum exercitationum lib. XV. De subtilitate ad Hieronymum Cardanum.

Francofurti: Apud Andream Wechelum, 1576, [16], 1130, [90] pp., illus., 17 cm. (8vo)


**Schaeffer,** Jacob Christian [1717 – 90]

Elementa entomologica.

Regensburg: Gedruckt mit Weissischen Schriften, 1766, [166] pp., 137 pls. (col.), 28 cm. (4to)


**Schwenckfeld,** Caspar [1563 – 1609]

Theriotropheum Silesiae, in quo animalium, hoc est quadrupedum, reptilium, avium, piscium, insectorum natura.

Lignicii: Impensis Davidis Alberti, 1603, [24], 564, [3] pp., 20 cm. (4to)


**Seager,** H. W.

Natural history in Shakespeare's time: being extracts illustrative of the subject as he knew it.

London: Elliot Stock, 1896, viii, 358 pp., illus., 22 cm.


**Sepp,** Christiaan *and* Jan Christiaan **Sepp** [1739 – 1811]

Beschouwing der Wonderen Gods, in de minstgeachtste Schepzelen. Of Nederlandische Insecten.

Amsterdam: J. C. Sepp, 1762 – 1821, 4 v., fronts. (col.), 200 pls. (col.), 25 cm. (small fol.)


**Serres,** d'Olivier de [1539 – 1619]

The perfect use of silke-wormes and their benefit. [Trans.] by Nicholas Geffe, with an annexed discourse of his own.

London: Felix Kyngston, 1607, [6], 100, 216 pp., illus., 19 cm. (4to)


**Simmler,** Josias [1530 – 76]

Vita clarissimi philosophi et medici excellentissimi Conradi Gesneri . . .

Tiguri: excudebat Froschoverus, 1566, 52 leaves, illus., 20 cm. (4to)

**Singer,** Charles

Greek biology and Greek medicine.

Oxford: Clarendon P., 1922, 128 pp., illus., 6 pls., 19 cm.


**Singer,** Charles

The herbal in antiquity and its transmission to later ages.

(Repr. from ''Jnl. Hellenic Studies'', vol. 47, 1927, pp. 1 – 52, illus.,
10 pls. (col.))


**Solmi,** Arrigo

The making of modern Italy.

London: Benn, 1925, xxiii, 231 pp., front. (map), illus., 22 cm.


**Le Spectacle de Nature**

Le spectacle de nature ou entretiens sur les particularités de l'histoire naturelle,
vol. 1. 3rd edition.

Paris: Chez La Veuve Estienne, 1735, xxiv, 562 pp., front., 24 pls., 18 cm.
(12mo)


**Spence,** William *jt. auth.*

An introduction to entomology.

*See*
Kirby, William *and* William Spence


**Sprague,** T. A.

The herbal of Otto Brunfels.

(Repr. from ''Jnl. of the Linnean Society, Botany'', vol. 48, 1928, pp. 79 – 124)


**Sprague,** T. A. *and* E. Nelmes

The herbal of Leonhart Fuchs.

(Repr. from ''Jnl. of the Linnean Society, Botany'', vol. 48, 1931, pp. 545 – 642)

**Sprague,** T. A. *and* M. S. **Sprague**

The herbal of Valerius Cordus.

(Repr. from "Jnl. of the Linnean Society, Botany", vol. 52, 1939, pp. 1 – 113)


**Stelluti,** Francesco [1577 – 1652]

Persio tradotto in verso sciolto e dichiarato da Francesco Stelluti.

Rome: [s.n.], 1630, [24], 218, [20] pp., front., illus., 23 cm. (4to)


**Sulzer,** Johann Heinrich [1735 – 1813]

Die Kennzeichen der Insekten.

Zurich: [s.n.], 1761, xxviii, 204, 68 pp., 24 pls. (col.), 23 cm. (4to)


**Swammerdam,** Jan [1637 – 80]

Ephemeri vita: or the natural history and anatomy of the ephemeron, a fly that lives but five hours.

London: Printed for Henry Faithorne, 1681, [8], 44, [8] pp., 4 pls., 22 cm. (4to)


**Swammerdam,** Jan

Histoire generale des insectes.

Autrecht: Chez Guillaume de Walcheren, 1682, [8], 215 pp., 13 pls., 21 cm. (4to)


**Swammerdam,** Jan

Historia insectorum generalis.

Lugd. Batavorum: Apud Jordanum Luchtmans, 1685, [20], 212, [17] pp., 13 pls., 21 cm. (4to)


**Swammerdam,** Jan

Historia insectorum generalis.

Ultrajecti: Ex officina Otthonis de Vries, 1693, [18], 212, [17] pp., 13 pls., 21 cm. (4to)

**Swammerdam,** Jan

The book of nature: or, the history of insects. Translated from the Dutch and Latin by Thomas Flloyd, revised . . . by John Hill.

London: Printed for C. G. Seyffert, 1758, [4], xxviii, 236, 153, lxiii, [12] pp., 53 pls., 43 cm. (fol.)


**Theobald,** *Bishop*

Physiologus: a metrical bestiary of twelve chapters, trans. by Alan Wood Rendell.

London: Bumpus, 1928, xxvii, [34], 100 pp., front., 14 pls., 21 cm.


**Theophrastus** [c. 372 – c. 287 B.C.]

Enquiry into plants and minor works on odours and weather signs, trans. by Sir Arthur Hort.

London: Heinemann, 1916, 2 v., 17 cm.


**Thomson,** James

Archives entomologiques ou recueil contenant des illustrations d'insectes nouveaux ou rares.

Paris: Soc. Entomologique de France, 1857 – 8, 2 v., 28 pls. (col.) + 7 pls., 27 cm.


**Thorley,** John

Melisselogia. Or, the female monarchy.

London: Printed for the author, 1744, xlvi, 206, [2] pp., front., 4 pls., 21 cm. (8vo)


**Topsell,** Edward

The Historie of serpents. Or, the second book of living creatures.

London: Printed by William Jaggard, 1608, [10], 315, [7] pp., illus., 32 cm. (fol.).

[Missing pp. 99 – 102, 187 – 92, 241 – 52]

**Topsell,** Edward

The history of four-footed beasts and serpents.

London: Printed by E. Cates, 1658, bound in 2 v., illus., 33 cm. (fol.)


**Tragus,** Hieronymus

*See*
Bock, Hieronymus


**Turner,** William [c. 1515 – 68]

Libellus de re herbaria novus.

Londini: Apud Ioannem Byddelum, 1538, [20] pp., 22 cm. (4to)

[Copy made from an original edition]


**Turner,** William

The names of herbes, ed. by James Britten.

London: English Dialect Society, 1881, vii, 134 pp., 22 cm.


**Tusser,** Thomas [?1524 – 1580]

Five hundred pointes of good husbandrie, ed. by W. Payner and S. J. Herrtage.

London: English Dialect Society, 1878, xxxii, 350 pp., 24 cm.


**Venn,** John

John Caius . . .: a biographical sketch.

Cambridge: C.U.P., 1910, [ii], 78 pp., front., 7 pls., 24 cm.


**Vettori,** Pietro (Petrus Victorius) [1499 – 1585]

Explicationes suarum in catonem, varronem Columellam castigationum.

Paris: Robert Stephan, 1543, 72 leaves, 18 cm. (8vo)


**Vida,** Marcus Hieronymus [1470 – 1566]

Scacchia, ludus: a poem on the game of chess. Translated into English verse by the Rev. Samuel Pullein.

Dublin: Printed by S. Powell, 1750, [4], 95 pp., illus., 23 cm. (4to)

**Vida,** Marcus Hieronymus

The silkworm: a poem. In two books. Translated into English verse by the Reverend Samuel Pullein.

Dublin: Printed by S.Powell, 1750, 141, [1] pp., front., 23 cm. (4to)


**Virgil** [70 – 19 B.C.]

The works of Virgil, trans. by Davidson, revised by T. A. Buckley.

London: Bell, 1883, ix, 404 pp., front., 19 cm.


**Virgil**

The Georgics of Virgil. Trans. by C. Day Lewis.

London: Cape, 1941, 95 pp., 23 cm.


**Warder,** Joseph [fl. 1688 – 1718]

The true amazons: or, the monarchy of bees. 3rd edition.

London: Printed for John Pemberton, 1716, [14], 120 pp., illus., 19 cm. (8vo)


**Warder,** Joseph

The true amazons: or, the monarchy of bees. 7th edition.

London: Printed for T. Longman, 1742, 164 pp., front., illus., 18 cm. (12mo)


**Warder,** Joseph

The true amazons: or, the monarchy of bees. 8th edition.

London: Printed for T. Longman, 1749, 164 pp., front., illus., 18 cm. (12mo)


**Watkins,** M. G.

Gleanings from the natural history of the ancients.

London: Elliot Stock, 1885, xiii, 258 pp., 23 cm.


**White,** Stephen

Collateral bee-boxes, or, a new, easy and advantageous method of managing bees. 2nd edition.

London: Printed for L. Davies & C. Reymers, 1759, [2], 67 pp., front., 20 cm. (4to)

## White, W.

A complete guide to the mystery and management of bees.

London: Printed for the author, [1771], xvi, 94 pp., front., 21 cm. (8vo)

## Wildman, Thomas

A treatise on the management of bees; wherein is contained the natural history of those insects. 3rd edition.

London: Printed for W. Strahan, 1778, xx, 326, 16 pp., 3 pls., 22 cm. (8vo)

## Wilkes, Benjamin [? – 1749]

Twelve new designs of English butterflies.

London: the author, 1742, [1] p., 11 pls., 37 cm. (fol.)

[Page detailing means of forming a butterfly collection and pl. 9 missing]

## Wilkes, Benjamin

Twelve new designs of English butterflies.

London: the author, 1742, [2] pp., 12 pls. (col.), 32 × 41 cm. (fol.)

## Wilkes, Benjamin

The English moths and butterflies: together with plants, flowers and fruits whereon they feed, and are usually found.

London: the author, [c.1749], [24], 63, [4] pp., 120 pls. (col.), 33 cm. (fol.)

## Willughby, Francis [1635 – 72]

Ornithologiae libri tres.

Londini: Impensis Joannis Martyn, 1676, [10], 307, [5] pp., 75 pls., 37 cm. (fol.)

## Willughby, Francis

Ornithologiae libri tres.

Londini: Impensis Joannis Martyn, 1676, [10], 307, [5] pp., 31 pls., 37 cm. (fol.)

[Missing numerous plates; this copy would appear to be one of the few first issues surviving with its black & red t.p., and coat of arms]

**Willughby,** Francis

The ornithology of Francis Willughby. In three books . . . Translated into English, and enlarged with many additions . . . by John Ray.

London: Printed by A. C. for John Martyn, 1678, [12], 441, [6]., 80 pls., 37 cm. (fol.)


**Willughby,** Francis

De historia piscium libri quatuor.

Oxonii: E theatro Sheldoniano, 1686, [9], 343, [56] pp., 187 pls., 38 cm. (fol.)


**Worlidge,** J.

Systema agriculturae; being the mystery of husbandry discovered.

[London]: Printed for Tho. Dring, 1675, [36], 324, [4] pp., 31 cm. (4to)


**Worlidge,** John

Apiarium; or a discourse of bees.

London: Printed by Thomas Dring, 1676, [8], 28 pp., front., 17 cm. (8vo)


**Worlidge,** John

Apiarium; or a discourse of the government and ordering of bees. 2nd edition.

London: Printed for Thomas Dring, 1678, [8], 42, [12], pp., front., 19 cm. (4to)


**Worlidge,** J.

Vinetum Britannicum: or a treatise of cider, and other wines and drinks extracted from fruits growing in this kingdom. 2nd impression, much enlarged.

London: Printed for Thomas Dring, 1678, [22], 240 pp., front., illus., 1 pl., 19 cm. (4to)


**Wotton,** Edward [1492 – 1555]

De differentiis animalium libri decem.

Lutetiae Parisiorum: Apud Vascosanum, 1552, [12], 220, [13] leaves, 36 cm. (fol.)


**Wright,** Thomas, *ed.*

Popular treatises on science written during the middle ages.

London: Historical Society of Science, 1841, xvi, 140 pp., 23 cm.